A Picture Book of

UNDERWATER LIFE

Written by Theresa Grace
Illustrated by Roseanna Pistolesi

Troll Associates

SEA TURTLE

Most turtles have a special hiding place—their shell! They tuck their head, legs, and tail in it when they are in danger. But the sea turtle, which lives in the ocean, cannot hide in its shell. Its long legs end in flippers instead of feet to help it swim fast.

Sea turtles lay their eggs on land. They bury them in the sand, then leave. Sea turtles lay more than 100 eggs in a group, or *clutch*. When the babies are born, they find their way to the water all by themselves!

Library of Congress Cataloging-in-Publication Data
Grace, Theresa.
   A picture book of underwater life / by Theresa Grace; illustrated by Roseanna Pistolesi.
      p.   cm.
   Summary: Brief text and illustrations introduce twelve aquatic creatures including the sea horse, octopus, and penguin.
   ISBN 0-8167-1906-3 (lib. bdg.)   ISBN 0-8167-1907-1 (pbk.)
   1. Marine fauna—Juvenile literature.   2. Aquatic animals—Juvenile literature.   [1. Marine animals.   2. Aquatic animals.]
I. Pistolesi, Roseanna, ill.   II. Title.
QL122.2.G73   1990
591.92—dc20   89-37330

Copyright © 1990 by Troll Associates, Mahwah, N.J.
All rights reserved. No portion of this book may be reproduced in any form, electronic or mechanical, including photocopying, recording, or information storage and retrieval systems, without prior written permission from the publisher. Printed in the U.S.A.

10   9   8   7   6   5   4   3   2   1

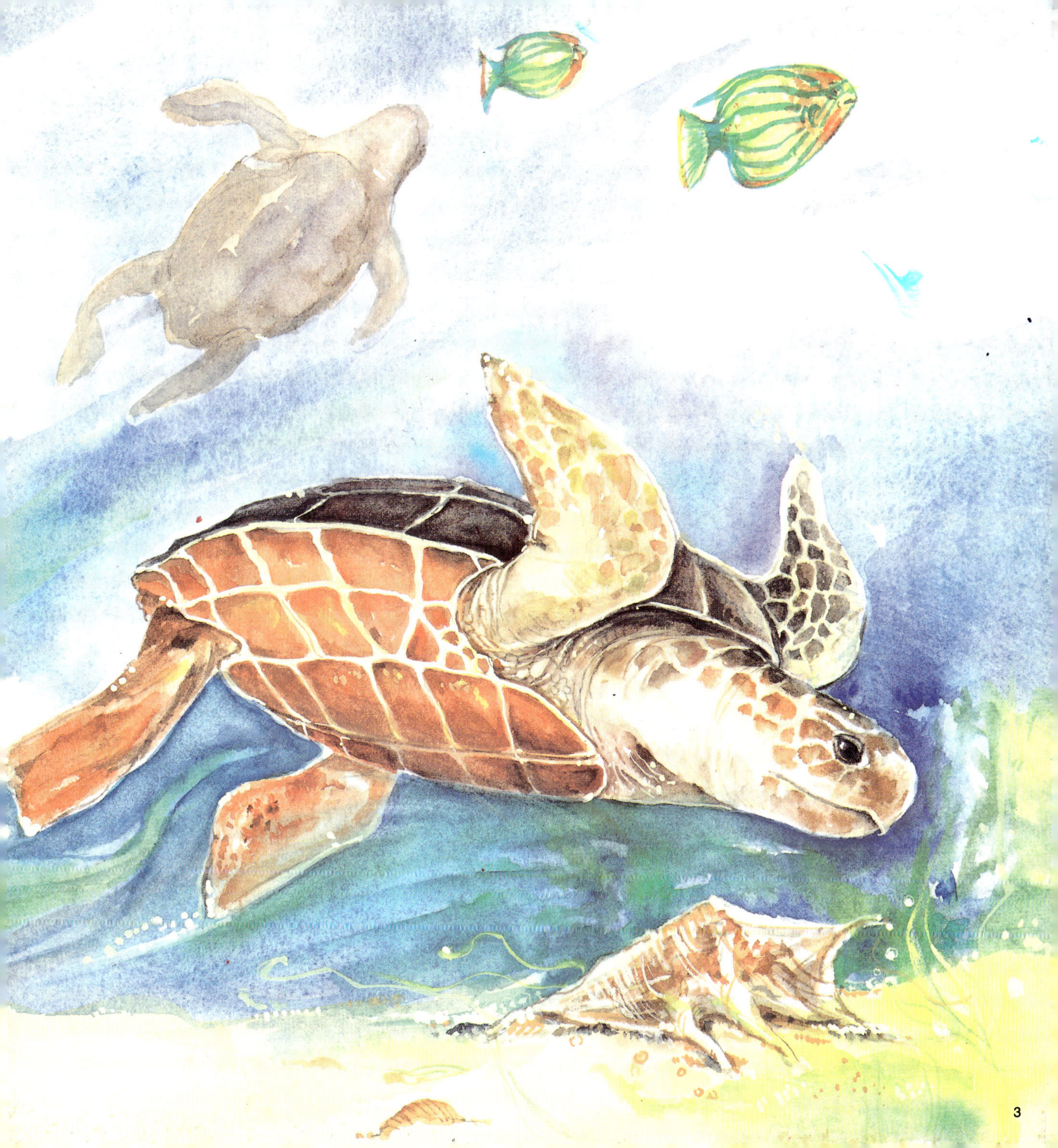

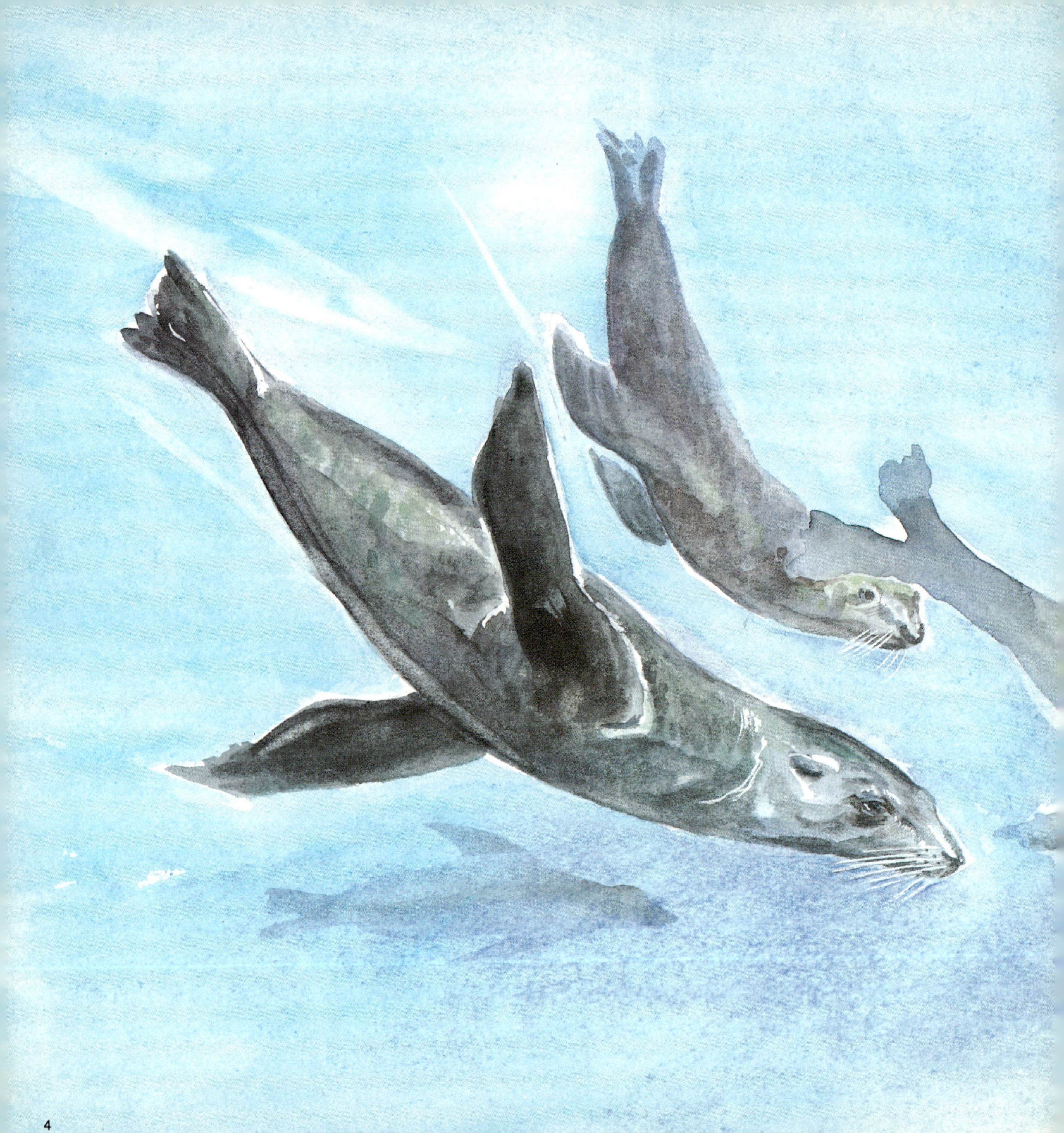

SEAL

In the Arctic and Antarctic, seals like to float on big pieces of ice and sleep on rocks. They walk on their flippers or slide on their bellies. A layer of fat, called *blubber*, keeps them nice and warm. Many seals have fur coats, too.

Most seals are great ocean swimmers. One kind of seal travels, or *migrates*, 5,000 miles every year. It never leaves the water during its long journey.

The seals that perform at zoos and circuses are sea lions. They can balance balls and do other tricks. And when the show is over, they can even clap for themselves!

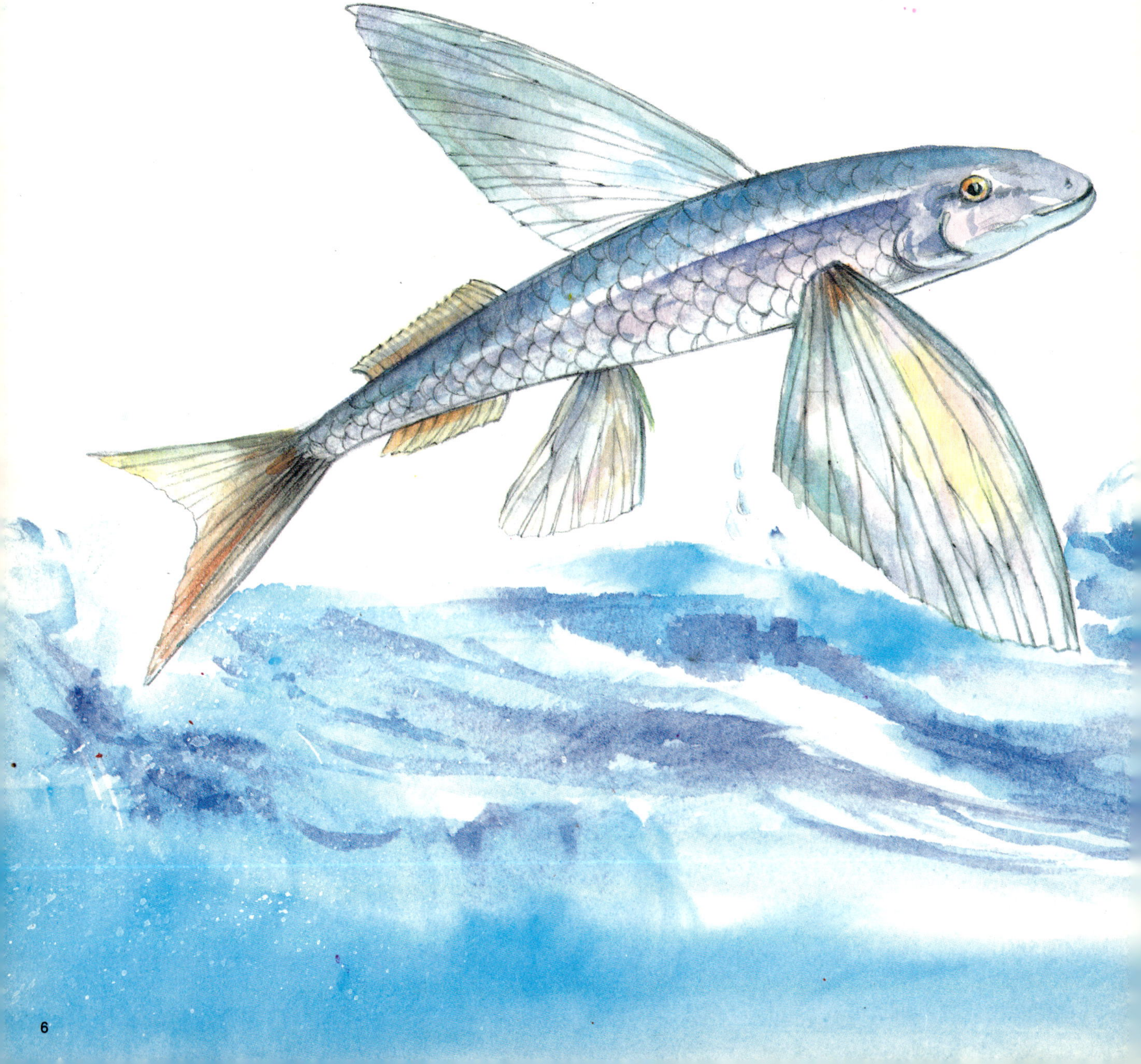

FLYING FISH

Flying fish swim in groups, called *schools*, in warm ocean waters. They cannot really fly. But they can *glide*. When there is danger below the surface, a flying fish's specially designed tail can lift its body out of the water. Then, the fish extends its long fins, which act like wings. It is gliding! Most trips are short ones—less than 200 feet. That's just far enough to get away from an underwater enemy.

LOBSTER

How would you like to wear armor all day long? The lobster does! Lobsters are covered with a tough, waterproof shell that protects them. Lobsters can also get away from their enemies by swimming—backwards!

A lobster's large, heavy claws help it catch and eat food. One claw has thick teeth. The lobster uses it to crush its prey. The other claw has sharp teeth to cut up the food.

Usually once every two years, a mother lobster lays 5,000-100,000 eggs. She carries them under her tail for about a year. Baby lobsters are very tiny. They swim on top of the water for about five weeks. Then they sink to the bottom, where they will live for the rest of their lives.

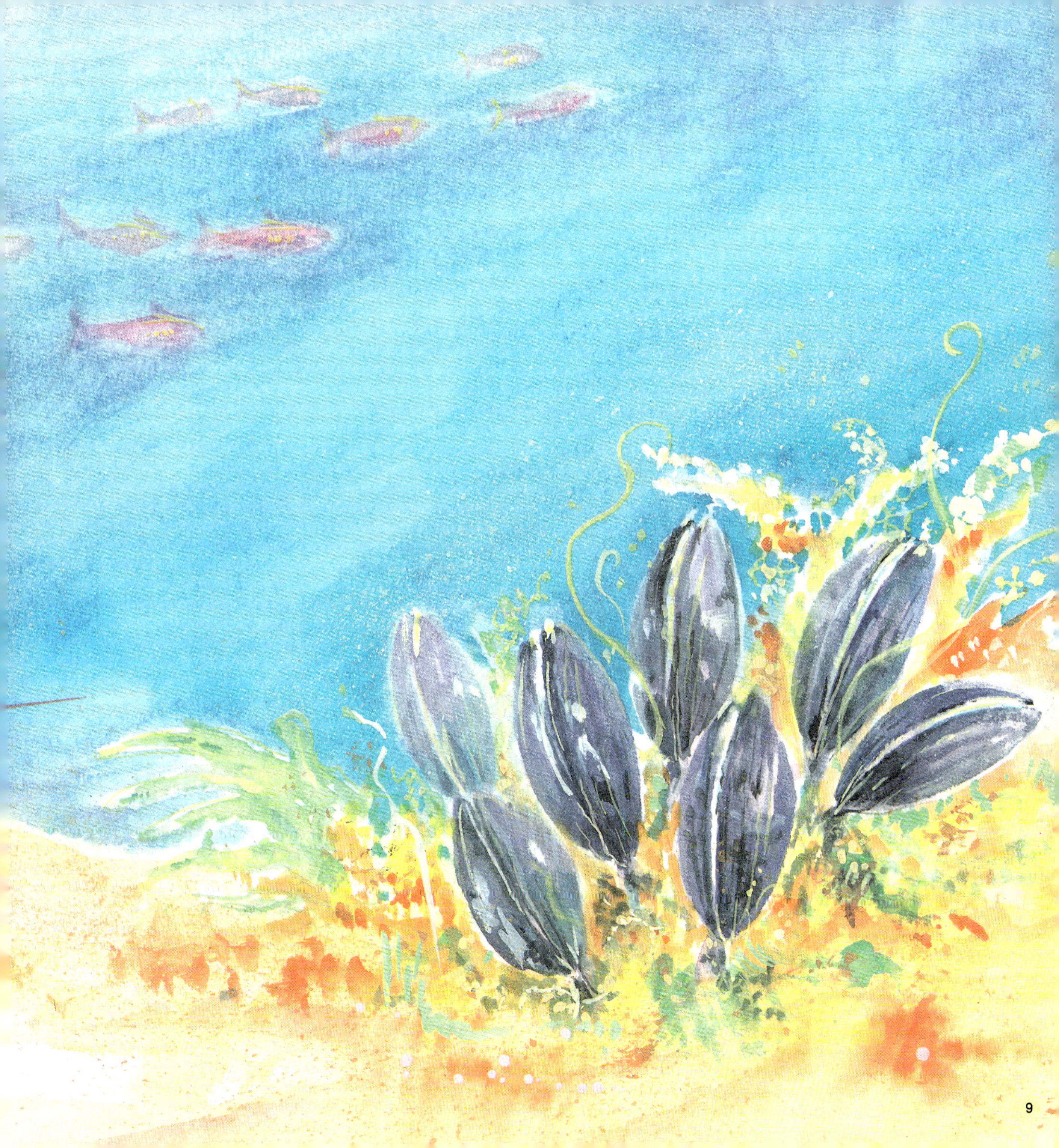

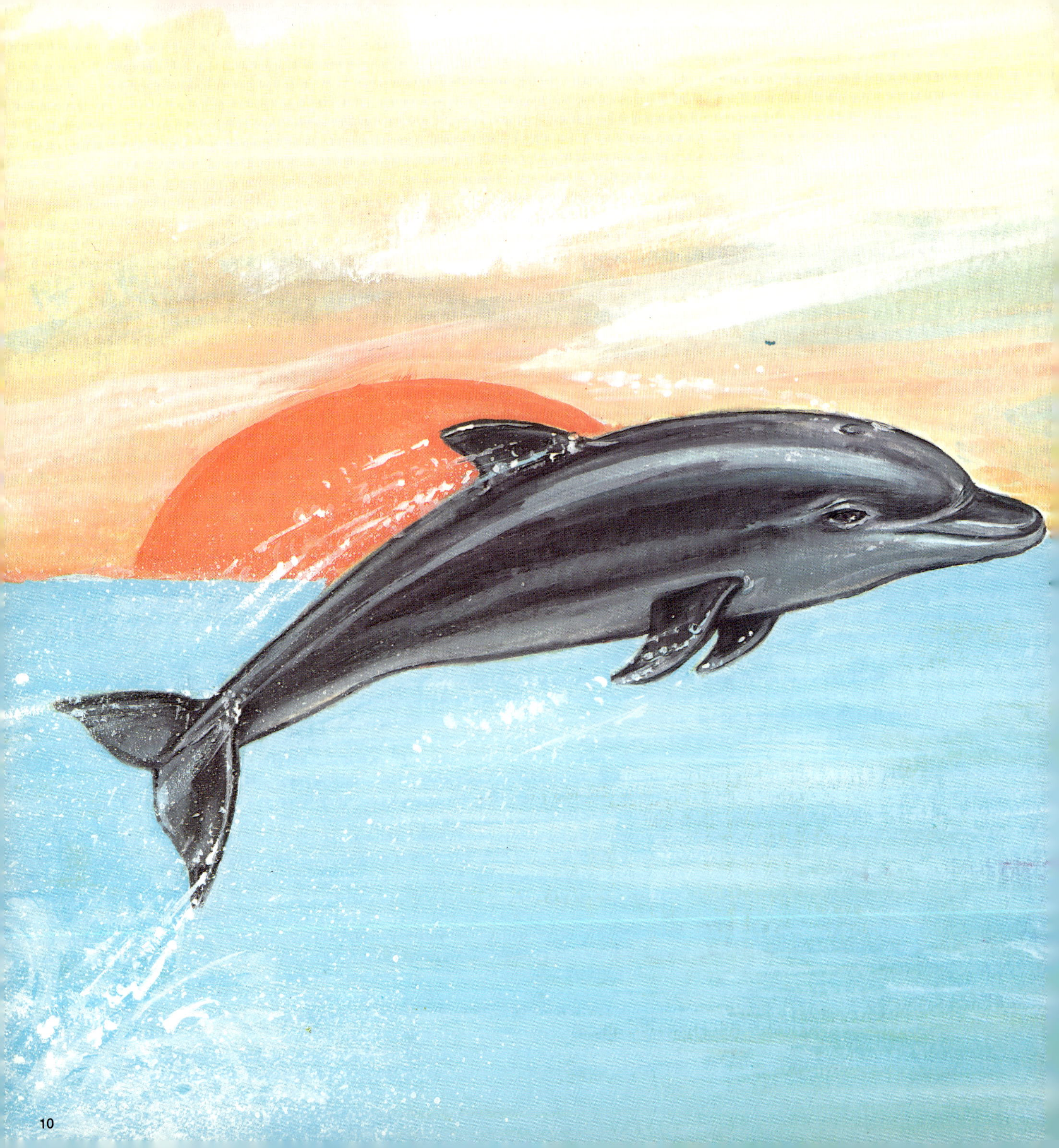

DOLPHIN

A dolphin may look like a fish and live in water, but it isn't a fish at all. It is a *mammal*. This means it is warm-blooded and breathes air. A dolphin swims to the surface of the water and breathes through a *blowhole* on top of its head. The blowhole closes when the dolphin is underwater, so no water gets in.

A dolphin makes clicking and whistling sounds. By listening to the way the sounds bounce off objects, dolphins can tell where things are in the water. Dolphins also use sounds to "talk" to one another.

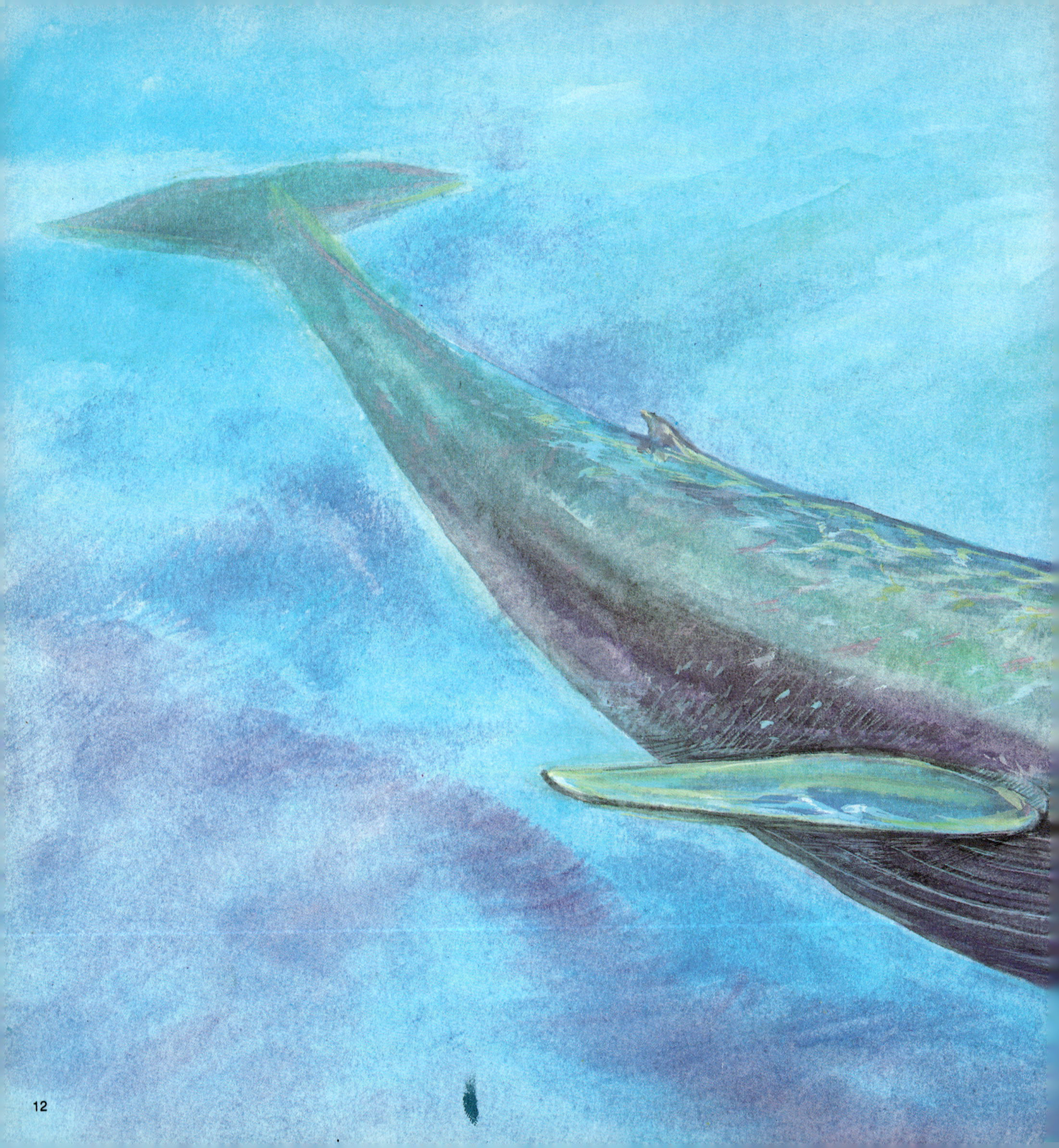

BLUE WHALE

The blue whale is probably the largest animal that ever lived. It weighs as much as 18 elephants and is longer than a basketball court. Yet this huge animal eats very tiny food. Instead of teeth, blue whales have bony plates called *baleen* in their mouths. The baleen strain tiny shrimplike animals called *krill* and even smaller plants and animals called *plankton* from the water. A blue whale can eat 11 tons of food in one meal!

A blue whale is a mammal. It comes to the surface to breathe through its blowhole. Then it dives deep under the water. It can stay underwater for almost an hour!

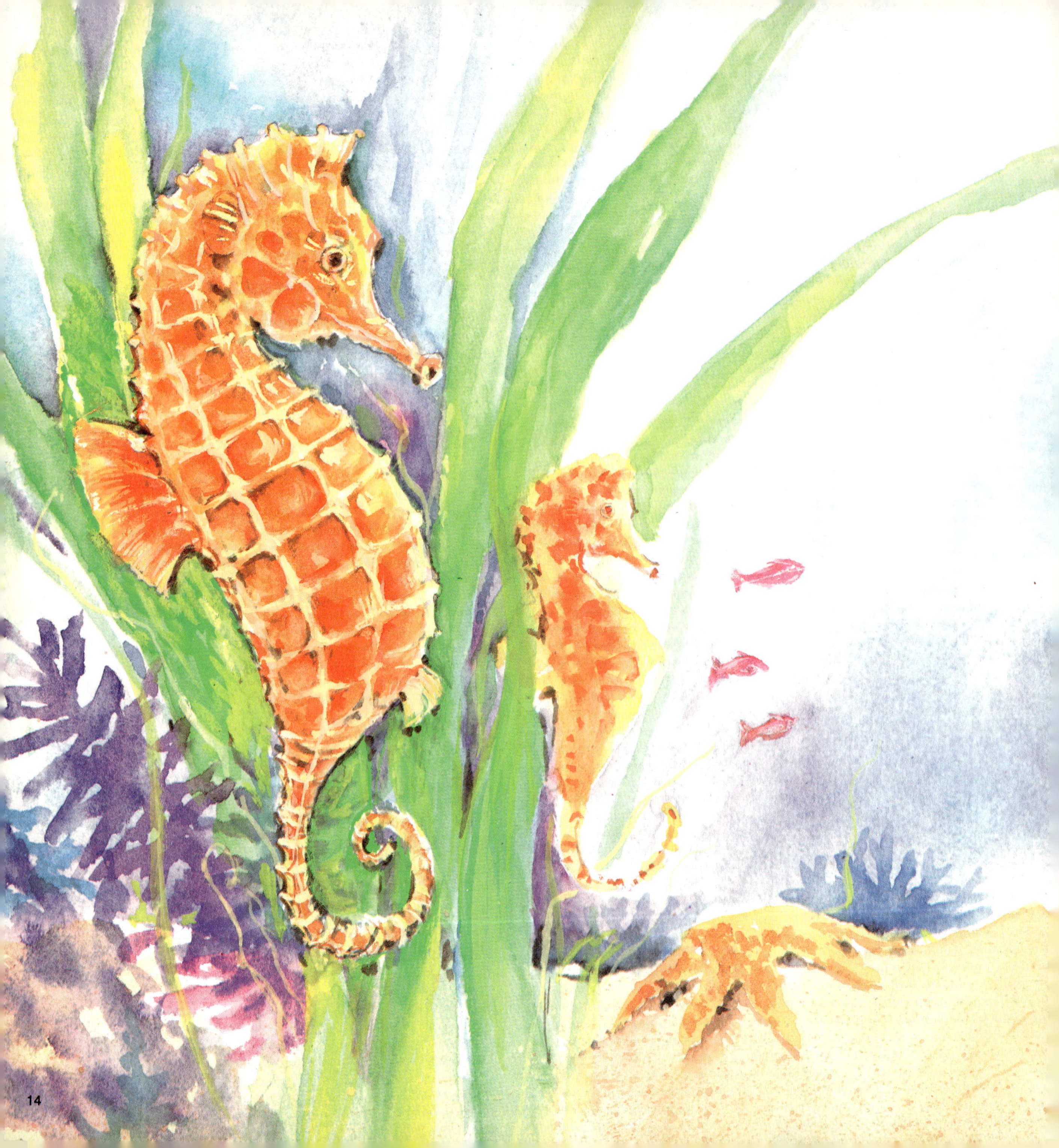

SEA HORSE

There is a horse that lives underwater—but it isn't a horse at all! The sea horse is really a fish. It got its name because its head looks like a horse's head.

Female sea horses lay about 200 eggs at a time. But it is the male who carries them until they hatch. He keeps the eggs in a pouch on the bottom of his body.

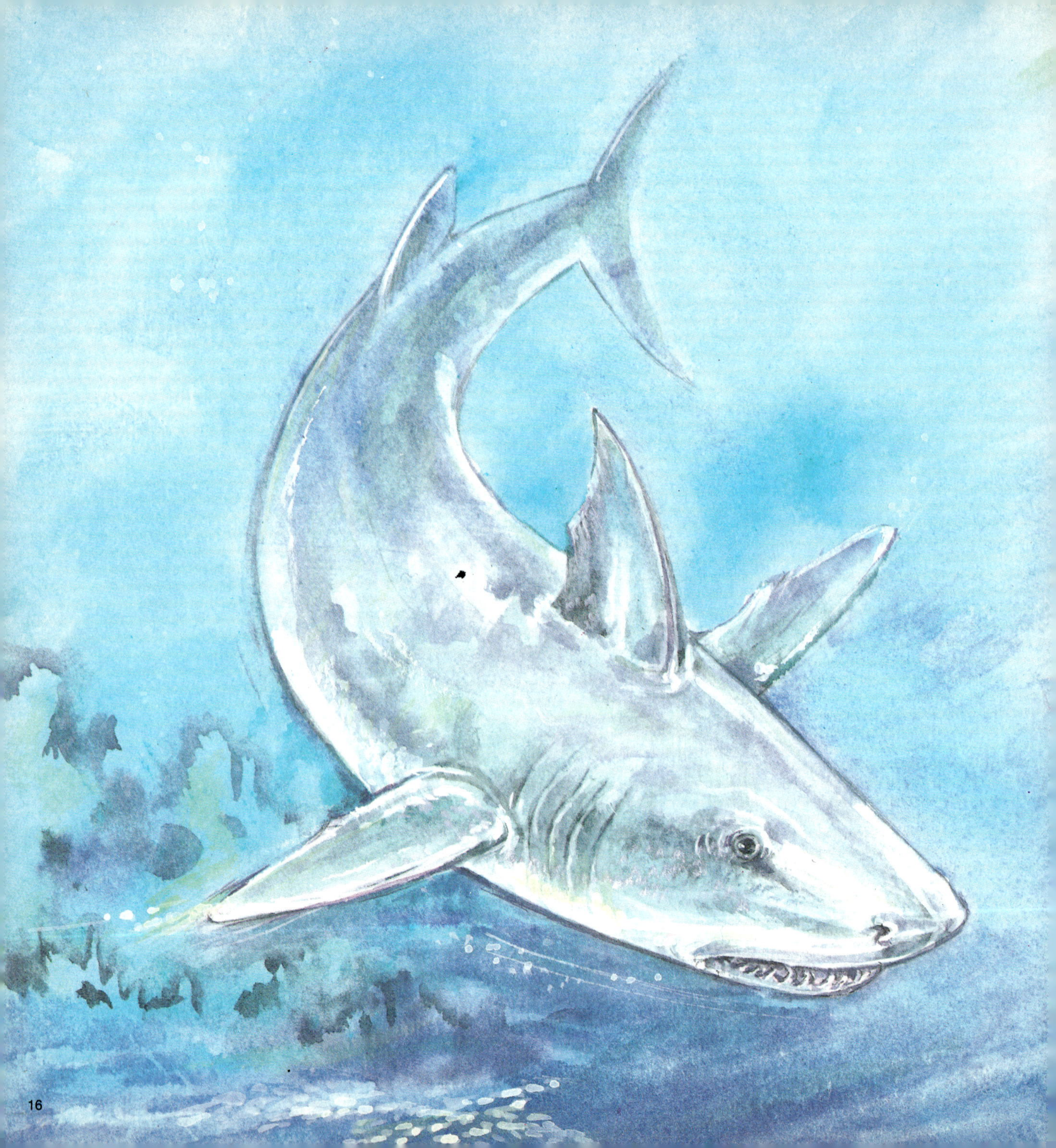

SHARK

There are many different kinds of sharks. Some are small—less than a foot long. Others are 60 feet long. The best known is the white shark. This shark is very fierce. It will attack and eat anything—even people.

A shark has many rows of sharp teeth. If it loses one of its teeth, another one moves up to take its place.

A shark's skeleton has no bones. It is made of tough, elastic *cartilage.* Sharks have very good senses of smell and hearing. This helps them find things to eat. And sharks have little trouble catching their prey. They can swim over 40 miles an hour!

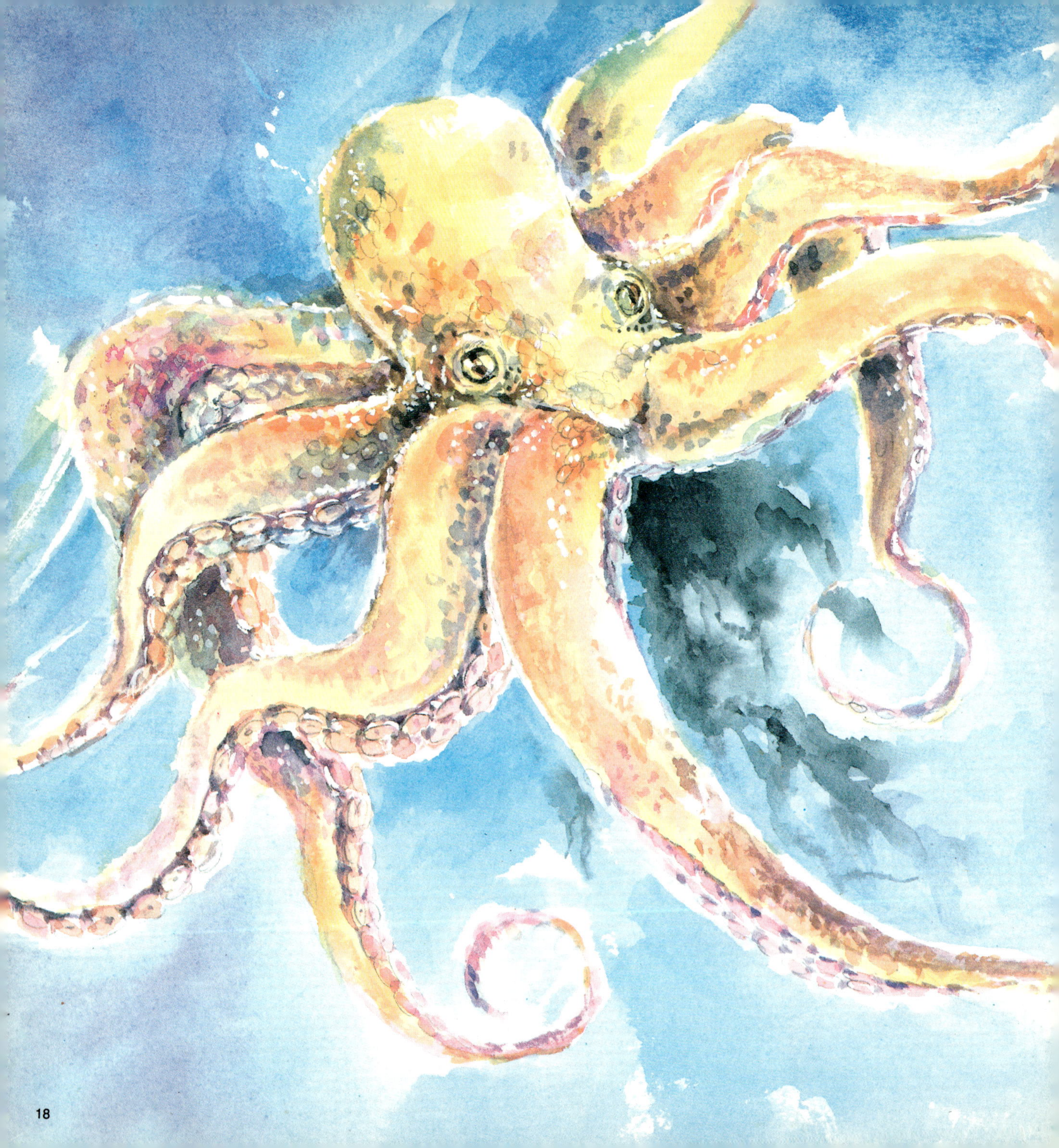

OCTOPUS

"Octopus" means "eight feet" in Greek. Each of the octopus' long *tentacles* has many round muscles called *suckers*. These help the octopus catch its favorite foods—clams, lobsters, and crabs. If an octopus loses a tentacle, another one will grow in its place.

If an octopus wants to make a quick getaway, it sucks water into a *siphon*, or opening, under its head. Then it forces the water out. This pushes the octopus through the water very fast. The siphon can also squirt a black fluid that hides the octopus. An octopus also hides by changing color. It can turn blue, brown, red, or even striped!

MANTA RAY

Manta rays are fish, but they look very different from most fish! Their fins spread out to the sides, just like wings. And instead of bones, a manta's flat body is held together with cartilage.

Another way that manta rays are different from fish is how they have their babies. Most fish lay eggs. Not the manta! Her young are born alive.

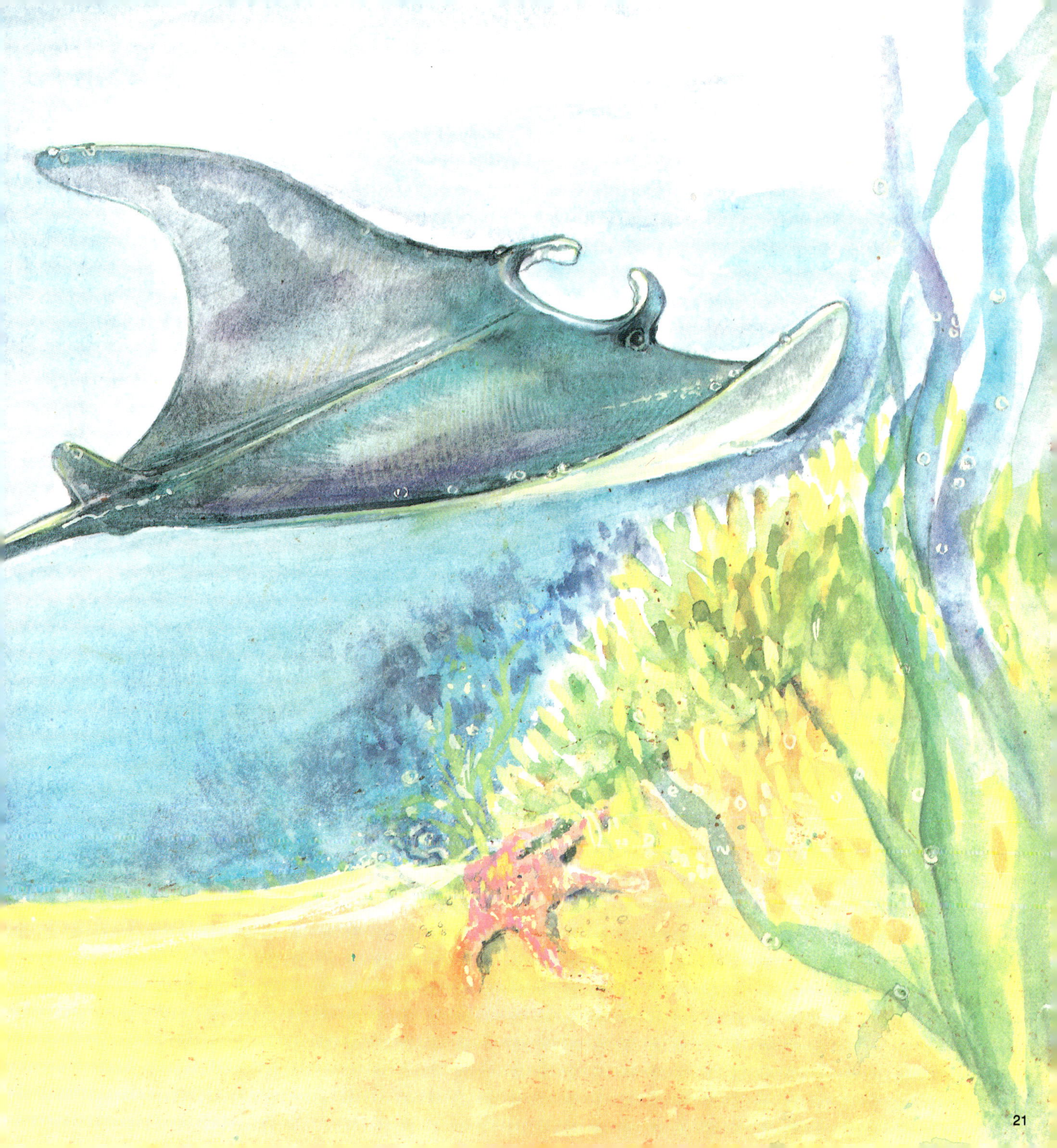

PENGUIN

A penguin has wings, but it doesn't use them for flying. It uses them to swim! By paddling with its wings and steering with its feet, a penguin can swim very fast.

Penguins live where it is very cold. Their feathers and a thick layer of fat help them keep warm as they swim and dive in the cold water. Penguins only come on land to breed and raise their chicks. A female penguin usually lays one egg. Her mate holds the egg between his feet and belly to keep it warm. After the chick is born, both parents take care of it until it is six months old.

23

FROG

Frogs start their lives as jellylike eggs in the water. When they are born, they are called *tadpoles*. They live in the water and breathe through gills, like a fish. But as frogs grow older, their bodies change. They grow legs and lungs. Now the frogs can breathe air and live on land. They like to sit near the water and catch insects with their long, sticky tongues.

A picture book of underwater life
J 591.92 Gra   16663

Grace, Theresa
    The Covenant School